Auroras

FIRE IN THE SKY

Published by Firefly Books Ltd. 2018

First paperback printing

Library of Congress Control Number: 2018937925

Library and Archives Canada Cataloguing in Publication
Bortolotti, Dan, author
Auroras : fire in the sky / Dan Bortolotti ; featuring photographs by Yuichi Takasaka.
Previously published: Richmond Hill, Ont. : Firefly Books, 2011.
Includes bibliographical references and index.
ISBN 978-0-228-10064-5 (softcover)
1. Auroras. 2. Auroras--Pictorial works. I. Takasaka, Yuichi, photographer II. Title.
QC971.B67 2018 538'.768 C2018-901691-4

Published in Canada by
Firefly Books Ltd.
50 Staples Avenue, Unit 1
Richmond Hill, Ontario L4B 0A7

Published in the United States by
Firefly Books (U.S.) Inc.
P.O. Box 1338, Ellicott Station
Buffalo, New York, USA 14205

Cover and interior design:
Janice McLean / Bookmakers Press Inc.

Printed in China

Canada *We acknowledge the financial support of the Government of Canada.*

Auroras

FIRE IN THE SKY

Dan Bortolotti

Featuring photographs by Yuichi Takasaka

FIREFLY BOOKS

A fish-eye lens captures the arc of an aurora over Prelude Lake in the Northwest Territories. The lights of the city of Yellowknife are visible on the left.

Contents

Night of the Crimson Sky

FOREWORD BY TERENCE DICKINSON

On the night of March 12-13, 1989, the most intense aurora display of the past half century painted the night sky over much of planet Earth into a swirling, kaleidoscopic light show. It was one of the most dramatic celestial events in modern history. Fortunately, the sky was perfectly clear that night from my backyard in rural eastern Ontario, where I took the photo reproduced here. The dominant red hue is typical of powerful all-sky auroras, as opposed to the pale green of most displays. At its peak, the strength of this sky show was extraordinary — bright enough to be visible through the light pollution in the downtowns of major cities. From my country backyard, the spectacle was fantastic.

The display began around 9 p.m. with pale green curtains and arcs in the north, typical of a standard aurora. But it wasn't long before it passed the routine level. By 10 p.m., the activity had escalated to blanket nearly the entire sky, despite the light of the crescent Moon. Rich red plumes emerged within the dancing auroral light that seemed to reach a crescendo at about 1:20 a.m.

The magnificent aurora of March 1989 was ignited by the most powerful magnetic storm in a generation. The vibrant red display was visible as far south as the Caribbean and northern Africa. Astronomer Terence Dickinson witnessed its glory from his observatory in eastern Ontario.

Just before the climax, as I was fiddling with my camera settings, I suddenly noticed the snow-covered ground brightening. Within about 15 seconds, the sky exploded into dazzling arcs, rays and curtains of blue, purple, yellow, green and, most impressive, brilliant red. The entire sky was awash in pulsating light. I rank this aurora just behind a total eclipse of the Sun in terms of visual impact and astronomical splendor. Throughout the entire display, there was seldom a time when fully three-quarters of the sky was without auroral glow.

What caused it? Two days before the memorable aurora, a huge flare was seen erupting on the Sun, dead center on the side facing Earth. Such eruptions result when intertwining magnetic fields on the solar surface toss into space a cloud of solar plasma made of electrically charged atomic particles. The particle cloud reached Earth two days later, became trapped by the Earth's magnetic field and was funneled to the polar regions, where interactions with air molecules produced the worldwide auroral lights. It was a global event, unprecedented in modern times. The aurora was seen as far south as Cuba and northern Africa.

But there was a dark underbelly to the cosmic light show. All this solar-induced electrical energy being dumped into the upper atmosphere overloaded the power grid in Canada's province of Quebec, and at 2:45 a.m. on March 13, the grid failed, leaving six million people without power in winter conditions for nine hours.

Although the Great Aurora of 1989 was visually stupendous, a few more recent displays have reached almost the same level, most notably the one on November 7, 2004, as shown in the photo on the facing page, also taken from my backyard. Looking back on more than 50 years of skywatching, I have seen the most active auroras during the months of March, April, September, October and November.

If you've ever been entranced by a spectacular aurora display — or especially if you haven't — you can savor the wonderful aurora portraits of Yuichi Takasaka, who has contributed many of the images that appear on the following pages. Writer Dan Bortolotti serves as a seasoned guide, charting human understanding of this powerful visual phenomenon as it is reflected in our cultural and scientific history.

Terence Dickinson is editor of SkyNews *magazine and author of the best-selling stargazing guidebook* NightWatch.

Forming a yin-yang of green and red, an auroral corona looms above the treetops in Terence Dickinson's rural backyard on November 7, 2004. This structure appears when an active aurora is directly above the observer, causing the parallel rays to appear as if they are radiating from a central point.

Nature's Light Show

WHERE SCIENCE MEETS SPLENDOR

Of all nature's visual spectacles, none uses a larger canvas than the aurora. On dark, clear nights, its ribbons unfurl over the entire dome of the sky, painting it with brushstrokes of green, yellow, pink and red.

The aurora terrified ancient peoples, who attributed the pulsating display to angry spirits or saw it as a harbinger of war. It inspired poets and philosophers in Europe and Asia, who glimpsed its beauty only when it stretched into the midlatitudes. The aurora was a beacon that guided the first polar explorers and a mystery that confounded scientists who struggled to explain its origin. Many of its secrets are still beyond our grasp.

We no longer believe that the aurora is supernatural. We know now its colors are emitted by atoms and molecules in the atmosphere that are excited by the charged particles of the solar wind. But as our understanding of nature's light show has evolved throughout the centuries, our wonder has remained undiminished.

A complex interaction between charged particles from the Sun, the magnetic field of Earth and the nitrogen and oxygen in the atmosphere produces the aurora, one of nature's most dynamic displays.

© CHRISTIE'S IMAGES/SUPERSTOCK

Aurora, the Roman goddess of the dawn, flew across the sky each morning to herald the arrival of her brother Helios, the Sun.

In 1619, Galileo wrote of "the sky at nighttime illuminated in its northern parts" and used the metaphor *borelae aurora*, or "northern dawn." Some 30 years later, French philosopher and scientist Pierre Gassendi adopted the term when describing a brilliant display he had seen in Europe, and aurora borealis became the common name for the phenomenon.

At right, a team of sled dogs navigates a northern dawn in Swedish Lapland.

© EMMANUEL BERTHIER/HEMIS/CORBIS

A ribbonlike aurora swirls over ships' masts in a northern British Columbia harbor, left. The auroral light show begins at the Sun, which throws off a stream of plasma — highly charged electrons and protons — called the solar wind. These particles hurtle toward Earth at up to 500 miles (800 km) per second, yet even at that speed, they take about 1½ days to reach Earth.

When the solar wind slams into the magnetic field surrounding Earth, some of the particles are directed toward the planet's magnetic poles — one in the north, one in the south. As the particles accelerate through the atmosphere, they collide with atoms and molecules of nitrogen and oxygen. The energy from the electrons is absorbed by these gases, then released as the colored light we call the aurora.

© PEKKA SAKKI/EPA/CORBIS

What we see as a shimmering curtain in the sky is actually an oval band that stretches around the Earth's magnetic poles. As the solar wind strikes Earth, the auroral ovals bulge slightly toward the equator on the nighttime side of the planet. The maximum expansion occurs near the midnight portion, which makes around midnight the best time to see an aurora.

Above, a towering display of the aurora borealis is seen over Finland. Because the north magnetic pole is in the Canadian Arctic, the aurora borealis is most often visible in Scandinavia (including Greenland), Alaska, northern Canada and northern Russia. Some magnetic storms are so powerful, however, that the auroral oval is pushed even farther south, bringing the aurora to the mid- and lower latitudes.

At right, a fish-eye lens accentuates the arc of an aurora, making it appear as though a rainbow is stretching from horizon to horizon.

Although the so-called northern lights occur around both of the planet's magnetic poles, they are only rarely witnessed in the less populous southern hemisphere. They would have been seen infrequently by the First Peoples of Australia, New Zealand, South Africa and Tierra del Fuego. Today, those who witness auroras over the south magnetic pole are likely to be stationed at remote Antarctic research stations, above, left and right.

Until the 18th century, Europeans were unaware that auroras occurred in the southern half of the globe. Captain James Cook wrote about his novel encounter during his second voyage to the southern hemisphere, in February 1773: "Lights were seen in the heavens, similar to those in the northern hemisphere, known by the name Aurora Borealis, or northern lights; but I never heard of the Aurora Australis being seen before."

Cook's term, which literally means "southern dawn," is now commonly used for auroras in the southern hemisphere.

Although we tend to call them all "neon signs," the colored lights in store windows contain several different gases. Neon gives off red light when it is exposed to electricity, while helium produces gold light, and argon turns green.

A similar principle is at work in the colors of the aurora. When the charged particles in the solar wind collide with gases in the Earth's atmosphere, the atoms and molecules become "excited" and emit light. Different gases at different altitudes glow with different colors.

The most common color is green, which is emitted by oxygen atoms up to about 200 miles (320 km) above the Earth's surface. Oxygen atoms higher in the atmosphere glow red. The purplish or pinkish glow along the bottom edge of an auroral curtain is produced by nitrogen molecules at a height of roughly 60 miles (100 km).

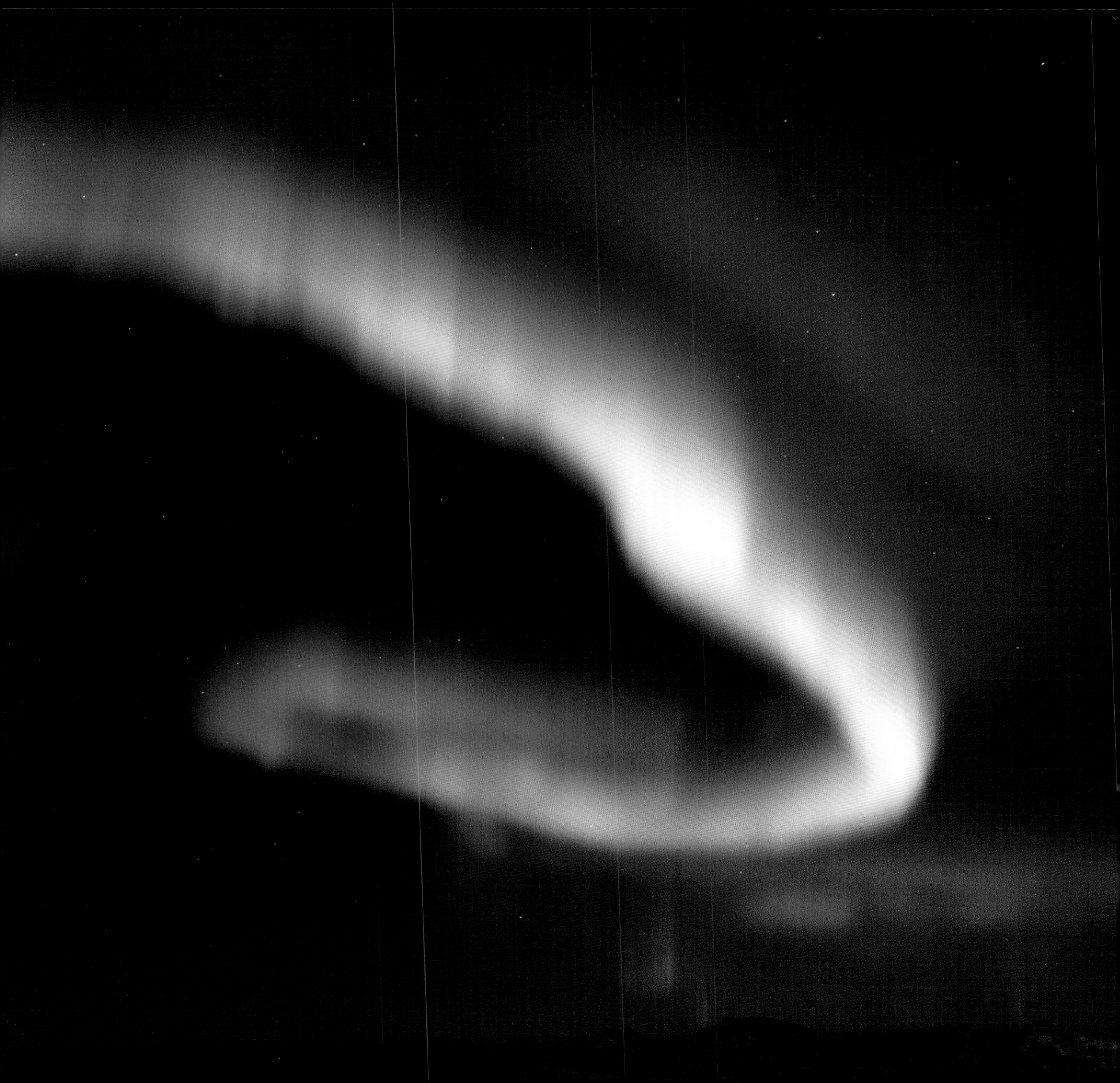

Red auroras follow on the heels of intense magnetic storms on the Sun. During these events, the entire sky may become blood-red as oxygen atoms are bombarded with massive amounts of energy at altitudes of hundreds of miles. Because of their magnitude, these displays can be visible at much lower latitudes than normal and, in extreme cases, may even be seen near the equator.

A crimson aurora is the Holy Grail of skywatchers. Photographer Yuichi Takasaka has observed more than a thousand auroras but has witnessed this extremely rare event only four times, including this one in Whitehorse, Yukon, in 2001.

© JENS MAYER/SHUTTERSTOCK

While the fiercest magnetic storms paint the entire sky a deep scarlet, less intense auroras include a mix of red, green and yellow hues.

This sequence of images taken over several minutes, near right, shows the aurora's changing form. Typically, an aurora appears as two to six distinct structures, becoming more diffuse as it moves overhead. The brightest section is usually at the bottom of the curtain, about 60 miles (100 km) above the Earth's surface. The formations are at the same altitude, but from the viewpoint of someone on the ground, they appear to radiate from a point on the distant horizon. The towering height of the curtain, which can extend well over 200 miles (320 km), is apparent as it swirls over Great Slave Lake in the Northwest Territories, far right.

Current Events

HOW MAGNETIC STORMS CAN WREAK HAVOC ON EARTH

A fiery red aurora blazes across the Alaskan sky, right. All-red auroras are as rare as they are beautiful. They occur when massive amounts of high-energy particles bombard oxygen atoms in the ionosphere at high altitudes. During powerful magnetic storms, the entire sky may become blood-red, a sight that often terrified ancient and medieval cultures.

The surface of the Sun is a violent place. Occasionally, the built-up magnetic energy in the Sun's atmosphere is expelled in a massive eruption of charged particles that are hurled into space. Think of it as a huge gust of solar wind.

It may take two to three days for this burst of energy to reach our atmosphere, but when it does, it unleashes a geomagnetic storm. This disruption in the Earth's magnetic field sometimes ignites exceptionally bright auroras and can cause electrical equipment to go haywire.

The first known geomagnetic storm was also the largest ever recorded. It occurred in late August and early September in 1859 and produced auroras so intense that they were seen in the Caribbean and Hawaii. According to one story from the period, miners in the Rocky Mountains awoke in the middle of the night thinking that the bright glow in the sky was the morning twilight. Telegraph operators were zapped with electric shocks — some even reported that they could send and receive messages only after they disconnected the batteries and ran on atmospheric current alone.

Almost a century and a half later — when humans had come to rely much more heavily on electricity — another geomagnetic storm bombarded Earth. In March 1989, a huge ejection from the Sun crippled the power grid in the province of Quebec, leaving some six million people in the dark. The storm also disrupted satellite communications, rendered compasses useless and caused automatic garage doors to open and close on their own. Meanwhile, brilliant auroras were visible as far south as Key West, Florida, and Mexico's Yucatán Peninsula.

Some scientists believe that if a geomagnetic storm as powerful as the 1859 event were to hit today, it might destroy low-orbiting satellites and cause continent-wide blackouts.

When an aurora stretches overhead, we see a spectacular multicolored display called a corona, above. The pinkish color along the bottom edge of the curtain is emitted by low-level nitrogen molecules. Viewed from many miles distant, the aurora assumes its more familiar structure, which takes on an added dimension when reflected in the water of a small pond, at right.

The aurora can take several forms: *arcs* follow a curved path from one horizon to the other; *rays* extend vertically like beams of light; *bands* have kinks or folds along their lower edges. When the auroral light covers a large expanse of the sky with little structure, it is called a *veil*.

Stories in the Sky

MYTHS AND LEGENDS OF THE AURORA

For as long as humans have gazed at the night sky, the aurora has been a source of mystery, delight and fear. The indigenous peoples of Alaska, northern Canada and Scandinavia all had legends about the origin of the dazzling light show they regularly witnessed during the long Arctic winter. In medieval Europe and other regions where auroras were only rarely visible, many people were terrified by what they saw. When the first polar explorers made expeditions to the top and bottom of the world, the colors of the aurora beckoned to them like beacons in the night.

Even without scientific traditions or technology, many cultures struggled to come up with natural explanations for the aurora. More often, they wove a tradition of myths and legends to give meaning to the lights. They saw them as the souls of their ancestors, the fires of distant tribes, the omens of war. Whether the aurora was regarded as comforting or threatening, however, it always made a deep impression on the people who lived beneath its glow.

The mysterious northern lights have ignited the imaginations of every culture that has lived under the auroral oval. At left, the lights shimmer above a Nisga'a totem pole in the northern British Columbia community of Gingolx.

An auroral glow illuminates Nisga'a Memorial Lava Bed Park in northern British Columbia. The park is the site of a volcanic eruption that occurred more than 250 years ago, destroying two Nisga'a villages and killing more than 2,000 people.

The Native peoples of Canada's Far North viewed the aurora with a combination of reverence and fear. Today, the Native-operated Aurora Village, located just outside Yellowknife, Northwest Territories, offers tourists an opportunity to experience the light show in an environment that evokes indigenous culture.

© GETTY IMAGES

Anthropologist Knud Rasmussen, right, visited many Inuit communities in the 1920s, during his 16-month journey by dogsled from the eastern Canadian Arctic to Alaska. According to one of the many legends he heard, spirits of the dead occasionally play soccer in the heavens, using a walrus skull as a ball. “It is this game of the souls playing at ball,” he reported, “that we see in the sky as the northern lights.”

Rasmussen also learned that many Inuit stories referred to whistling and crackling noises made by the aurora, a phenomenon scientists have never been able to document. “If one happens to be out alone at night when the aurora borealis is visible,” Rasmussen wrote of this myth, “one only has to whistle in return, and the light will come nearer out of curiosity.”

A 19th-century watercolor depicts the golden glow of an aurora over Norway, above. Norse mythology often associated the aurora with female warriors: One story held that the lights were caused by the Sun glinting off the armor and shields of the Valkyries.

In *The King's Mirror*, a Viking text written around 1250, a king and his son have a long discussion about the glow "that the Greenlanders call the northern lights." The father admits that he does not know their cause, but he shares several possible explanations. The lights may be the result of "fires that encircle the outer ocean," or perhaps "the frost and the glaciers have become so powerful that they are able to radiate forth these flames."

At right, an auroral band meanders above the snow-covered mountains near Kiruna, in the Far North of Sweden.

When an aurora lit up the skies over Bavaria in the 16th century, one artist's interpretation of the event, below, included armed warriors doing battle. Roughly translated, the inscription above the picture reads: "On December 28, 1560, the people of Bamberg and Lichtenfels witnessed a terrifying yet amazing sight."

While the aurora was familiar to people of the Far North, it was visible at midlatitudes only during the most powerful magnetic storms. Because these intense auroras are often blood-red, they frightened the inhabitants of medieval Europe, who considered them omens of war.

"The aurora borealis had become a source of terror," wrote a 19th-century French commentator. "Bloody lances, heads separated from the trunk and armies in conflict were clearly distinguished."

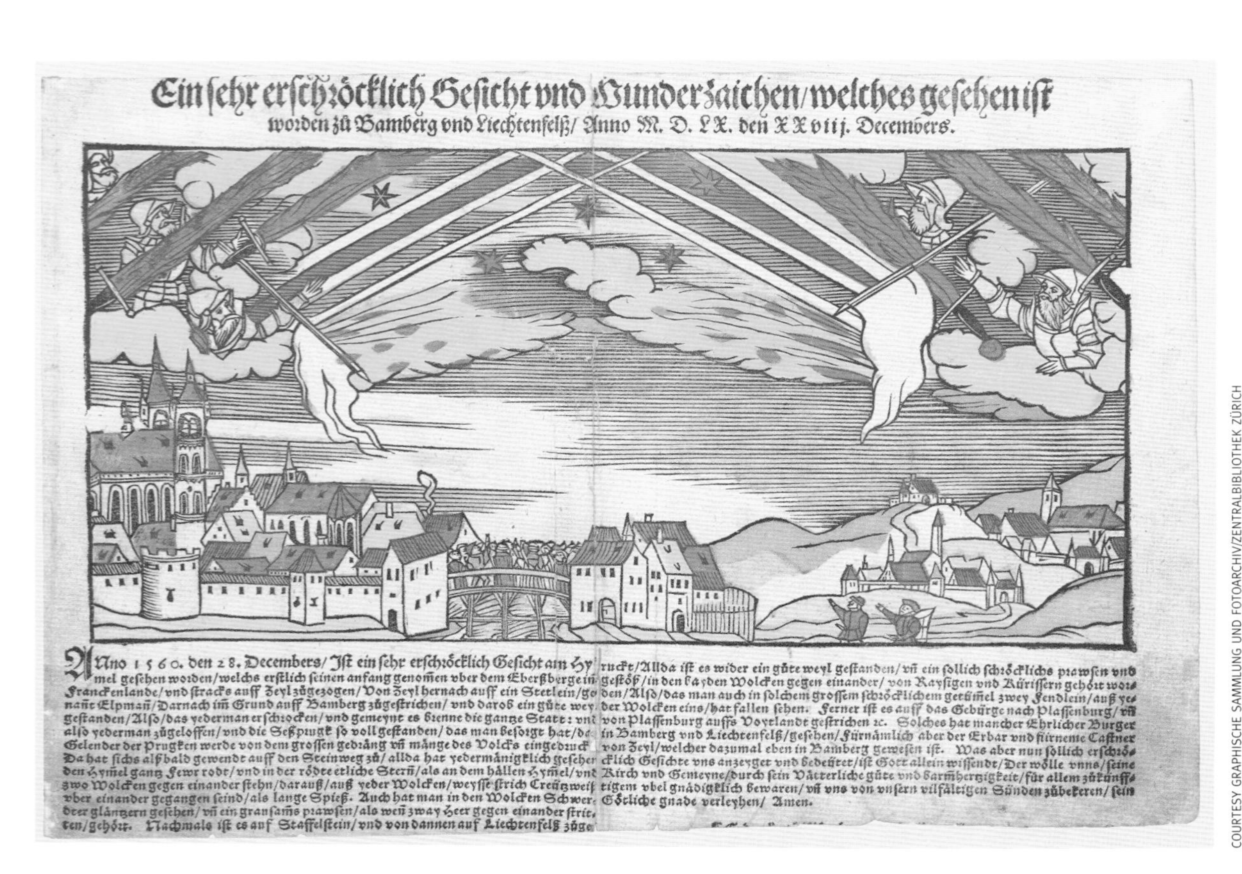

COURTESY GRAPHISCHE SAMMLUNG UND FOTOARCHIV/ZENTRALBIBLIOTHEK ZÜRICH

The North American Native peoples who lived at midlatitudes had their own stories about the dancing lights they occasionally saw in the North. The Makah of the Pacific Northwest have a long tradition of whaling. They believed that the aurora emanated from the fires of northern whaling communities, where blubber was being boiled in open pots.

The British explorers who traveled to the Canadian Arctic in the 19th century were captivated by the aurora. The sketch below was made at Lady Franklin Bay, Ellesmere Island, in the 1870s. Another of those who searched for the doomed expedition of Sir John Franklin wrote of the northern lights: “Who but God could execute them, painting the heavens in such infinite scenes of glory?”

When Norwegian explorer Fridtjof Nansen attempted to reach the North Pole in the 1890s, his ship became trapped in the pack ice. The aurora he observed from the deck helped dispel the gloom of the Arctic winter: “It was an endless phantasmagoria of sparkling color, surpassing anything that one can dream.”

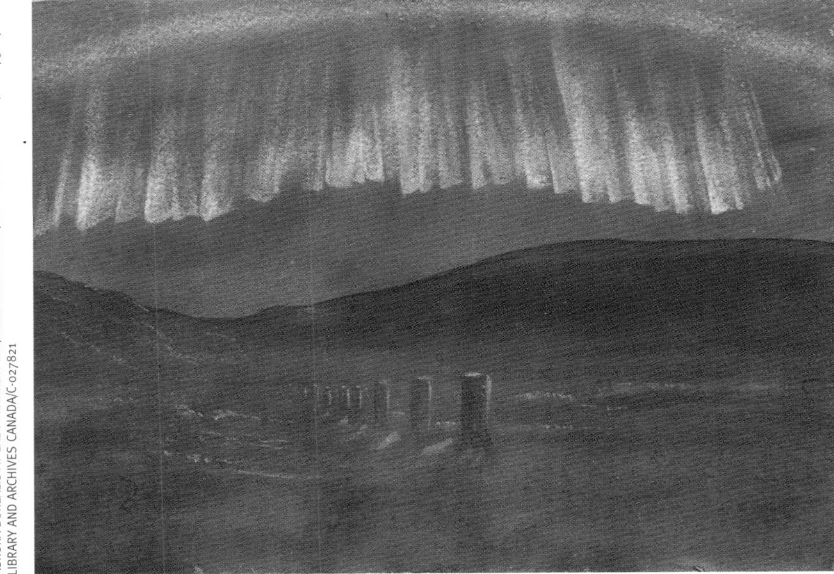

AURORA BOREALIS AT LADY FRANKLIN BAY, ELLESMERE ISLAND, BY THOMAS MITCHELL, C. 1875-1876.
LIBRARY AND ARCHIVES CANADA/C-027821

While it’s often repeated that the Japanese believe children conceived under the northern lights will enjoy prosperous lives, this cultural story is actually an urban legend. It may have arisen in northern Canada, where most of the aurora-watching tourists are Japanese. Even though auroras are extremely rare in Japan — Tokyo is at a lower latitude than San Francisco — the northern lights do have a prominent place in the country’s imagination. “Everybody in Japan knows what an aurora is,” says photographer Yuichi Takasaka. “We learn about them in school and on TV shows. They are everywhere.”

Night on the Marge

THE STORY OF SAM McGEE

The Northern Lights have seen queer sights,
But the queerest they ever did see
Was that night on the marge of Lake Lebarge
I cremated Sam McGee.

— Robert W. Service
"The Cremation of Sam McGee"

The sights are less queer today, but the northern lights can still be seen on the marge of Lake Laberge. The lake is a 30-mile-long (50 km) widening of the Yukon River, located north of Whitehorse.

Robert William Service was born in England in 1874 and immigrated to western Canada when he was 21. The poet and adventurer who came to be known as "the Bard of the Yukon" was actually a banker by trade. He worked for the Canadian Bank of Commerce in British Columbia until being transferred to the Yukon in 1904. Although he arrived after the Klondike Gold Rush had ended, Service was inspired by the stories of the miners and the hardships they'd endured.

In 1907, Service published his first collection of poetry, *Songs of a Sourdough*, which features "The Cremation of Sam McGee." The poem tells the tale of a prospector from Plumtree, Tennessee, who is overcome by the cold while "mushing [his] way over the Dawson trail." Realizing he won't survive the journey, McGee makes a last request of his partner: "I want you to swear that, foul or fair, you'll cremate my last remains."

When McGee dies, the narrator drags the corpse to a derelict ship that is stuck in the ice of Lake Laberge. (Service misspells the name in the poem, perhaps to emphasize the rhyme with "marge," which means "edge.") Fulfilling his promise to his friend, he builds a roaring fire in the old ship's boiler: "And I burrowed a hole in the glowing coal, and I stuffed in Sam McGee." The narrator later returns, expecting to find only charred remains. But when he opens the furnace door, he finds his grinning friend basking in the heat: "Since I left Plumtree, down in Tennessee, it's the first time I've been warm."

There really was a Sam McGee. He was a road builder from Ontario, and while he did journey to the Yukon, he didn't freeze to death. He was a client of the Whitehorse bank where Service worked. As the story goes, the banker-poet was working on a ballad and needed a name to rhyme with "Tennessee." McGee agreed to let Service use his name, unaware that he would have to endure jokes about the poem for the rest of his life. The real Sam McGee died of a heart attack in 1940 and was not cremated — he's buried in a small village northeast of Calgary, Alberta.

M. BRAVAIS, 1837-1838. FROM *THE AURORA BOREALIS* BY ALFRED ANGOT (NEW YORK, 1897). COURTESY SPECIAL COLLECTIONS, UNIVERSITY OF SASKATCHEWAN

PHOTOGRAPH OF DAWSON BY THE LIGHT OF THE AURORA BOREALIS, C. 1908. LIBRARY AND ARCHIVES CANADA/C-012125

While polar explorers like Fridtjof Nansen made sketches and woodcuts of the aurora, late-19th-century photographers repeatedly tried and failed to capture it with a camera. A few blurry, overexposed images were obtained, but none were of any scientific value. When French meteorologist Alfred Angot published his 1897 book *The Aurora Borealis*, he illustrated it with sketches like the one at left, top, of the curtainlike structures he described as "the most beautiful manifestations of the polar aurora."

The first auroral photographs were published around the turn of the 20th century. The image at left, bottom, shows a blazing aurora over the city of Dawson, Yukon, in 1908, the year after Robert Service published "The Cremation of Sam McGee." More than a century later, Yuichi Takasaka took an image of a towering auroral curtain, facing page, from just outside the same city.

The Southern Cross, the most familiar asterism in the southern hemisphere, hovers above an Antarctic aurora in this 1947 painting. On facing page, top, a 19th-century drawing depicts the aurora australis as seen from Melbourne, Australia, where the southern lights are rarely visible. The complete darkness of the Antarctic winter allows scientists on the continent many more opportunities to capture stunning auroral images, facing page, bottom.

FROM *THE AURORA BOREALIS* BY ALFRED ANGOT (NEW YORK, 1897).
COURTESY SPECIAL COLLECTIONS, UNIVERSITY OF SASKATCHEWAN

While the cultures of the Far North created a rich folklore about the polar lights, the indigenous peoples of the southern hemisphere have fewer legends, since they are seldom treated to the spectacle. The Maori of New Zealand believed the southern lights were the glow of fires lit by their ancestors, who had been carried away to the Southern Ocean in their canoes. Australian anthropologist Aldo Massola writes of one of the Aboriginal nations of the southeast: "When the Kurnai saw the aurora, they were thrown into great confusion, believing it to be the visible sign of the anger of the Great Man."

© KEITH VANDERLINDE/NATIONAL SCIENCE FOUNDATION

Seeing the Light

EARLY ATTEMPTS TO UNDERSTAND THE AURORA

More than 26 centuries ago, Greek philosophers began speculating about the origin of the rare and mysterious lights in the north. But they were stopped in their tracks in the fourth century BCE by Aristotle's pronouncement that the heavens were fixed and unchanging. Although this idea repeatedly clashed with what astronomers were observing in the skies, the extraordinary influence of Aristotle continued for some 2,000 years, and it was an obstacle to understanding what caused the aurora.

By the 18th century, scientists had noticed that the aurora was related to the Earth's magnetic field, and as they came to learn more about the relationship between electrical current and magnetism, it became clear that electricity is also involved. There were even clues that events on the Sun might play a role. However, although almost all the pieces were within reach, the puzzle of the northern lights remained unsolved as the 20th century opened.

As an aurora moves in over Great Slave Lake in Yellowknife, Northwest Territories, the rising Moon bathes the shore in golden light.

The prophet Ezekiel may have been the first to describe an aurora, in 592 BCE: "As I looked, a stormwind came from the north, a huge cloud with flashing fire, from the midst of which something gleamed like electrum."

Greek philosophers of the sixth century BCE may have witnessed the same aurora that Ezekiel saw, but they preferred more earthly explanations. Anaximenes of Miletus referred to "exhalations from the Earth," while Xenophanes of Colophon wrote of "burning clouds." Other Greek thinkers wondered whether the glow from the north was caused by reflected sunlight. Many centuries later, these same ideas were still being offered to explain the aurora. Even today, intense auroras, such as this one over Lake Wateree in South Carolina, have an otherworldly appearance.

COURTESY GRAPHISCHE SAMMLUNG UND FOTOARCHIVZENTRALBIBLIOTHEK ZÜRICH

In the sketch at left, residents of medieval Bavaria drop to their knees as an aurora paints the sky red. During the late Middle Ages, the influence of Aristotle dominated intellectual life in Europe. It took brave 16th-century thinkers like Copernicus and Galileo to challenge the prevailing ideas about Earth and the Sun and their relationship in the cosmos.

English scientist and physician William Gilbert was among those who supported Copernicus and his radical idea that Earth was not an immovable body around which the Sun revolved. In his masterwork *De Magnete* (*On the Magnet*), published in 1600, Gilbert suggested that Earth was a giant magnet, which explained why compass needles point north.

The lunar orb in these images adds to the spectacle of the northern lights, but in the 18th and 19th centuries, the Moon became yet another layer of the mystery. In 1789, English scientist John Dalton suggested that the aurora might be related to the tides, since his observations (wrongly) indicated that auroras occurred more frequently around the new and full phases.

Even Swedish physicist and Nobel Laureate Svante Arrhenius proposed that the Moon played a role. In the early 1900s, he argued that the lack of a lunar atmosphere would cause the Moon to become negatively charged by the solar wind and that auroras would be less frequent in any region where the Moon was above the horizon.

EDMUNDUS HALLEIUS R.S.S.
Astronomus Regius et Geometriæ Professor Savilianus

From 1645 until 1715, a period now called the Maunder Minimum, solar observers noted a surprising lack of sunspots. Although no one knew it at the time, this was the reason there were also remarkably few visible auroras during these seven decades. That all changed in March 1716, when Europe enjoyed its most brilliant auroral display in almost a century.

One of the fascinated observers was Edmond Halley, above, who recorded that the same display was seen across the continent and always to the north, which meant that the phenomenon was unlikely to have a local source, such as vapor escaping from the ground. His experiments eventually led him to recognize that auroras are related to the Earth's magnetic field.

FROM *THE AURORA BOREALIS* BY ALFRED ANGOT (NEW YORK, 1897). COURTESY SPECIAL COLLECTIONS, UNIVERSITY OF SASKATCHEWAN

© PHILADELPHIA MUSEUM OF ART/CORBIS

The above 19th-century sketch of an auroral curtain over Paris is neatly echoed in the photograph at right, taken in northern Canada more than a century later.

At a meeting of the Royal Academy of Sciences in the French capital in 1779, Benjamin Franklin proposed a theory about the aurora that evoked his famous experiment, left, when he flew a kite in a thunderstorm to demonstrate that lightning is a form of electricity. Franklin suggested that electricity in the clouds would be most concentrated over the Earth's poles and that it "must be condensed there and fall in snow, which electricity would enter the earth." If this were, indeed, true, he went on, "would it not give all the appearances of an aurora borealis?"

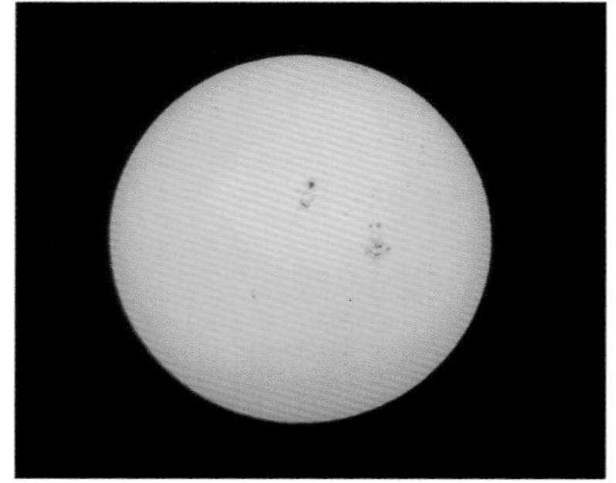
COURTESY TOM TSCHIDA/NASA

Sunspots, shown above, are regions of intense magnetic activity on the surface of the Sun. Ancient Chinese astronomers observed these dark irregularities as early as the fourth century BCE, and Galileo was among the first to study them with a telescope, in the early 1600s. By the mid-19th century, scientists had determined that sunspots appear and fade in a cycle that lasts about 11 years.

On September 1, 1859, English solar astronomer Richard Carrington became the first person to witness a solar flare. The bright flash erupted from a sunspot he was observing. Within 24 hours, Earth was blasted with a tremendous geomagnetic storm, which ignited brilliant auroras that were visible as far south as the Tropics. Yet even Carrington was not convinced that the two events were related. "One swallow does not make a summer," he wrote.

COURTESY SPECIAL COLLECTIONS, UNIVERSITY OF SASKATCHEWAN

All the subtle, three-dimensional structure of the aurora is evident in modern images, like the one at right, taken along the Ingraham Trail, in the Northwest Territories. Before the age of photography, artists and engravers did their best to depict the complicated forms of the aurora, and their illustrations were used in both scientific books and popular travelogues.

The sketch above shows a snake-like aurora undulating above a trading post in Nulato, Alaska. It appears in the 1868 book *Travel and Adventure in the Territory of Alaska*, by the intrepid British artist and adventurer Frederick Whymper.

The Milky Way sparkles in the northwestern sky near Whitehorse, Yukon. Scientists classify the brightness of an aurora using a scale from one (comparable to the Milky Way) to four (bright enough to cast shadows on the ground).

During the 18th century, prominent scientists continued to believe that the auroral glow was sunlight being reflected off particles high in the atmosphere. If that were the case, however, then it should be possible to use a prism to separate the light of an aurora into the familiar colors of the rainbow. In 1868, Swedish physicist Anders Jonas Ångström demonstrated that this is not the case, so the auroral light could not be sunlight.

Auroral rays appear to be streaming from the hilltops near Vee Lake, in the Northwest Territories. Perspectives like this make it easy to appreciate why one of the most difficult tasks for early scientists was measuring the aurora's altitude. Many were confident that the arcs and bands dipped very low in the sky — some believed that they occasionally made contact with the ground.

Even as late as 1910, the Encyclopaedia Britannica dramatically underestimated the height of the aurora, stating that some observers have seen it "below the clouds or between themselves and mountains."

© ALEX0001/SHUTTERSTOCK

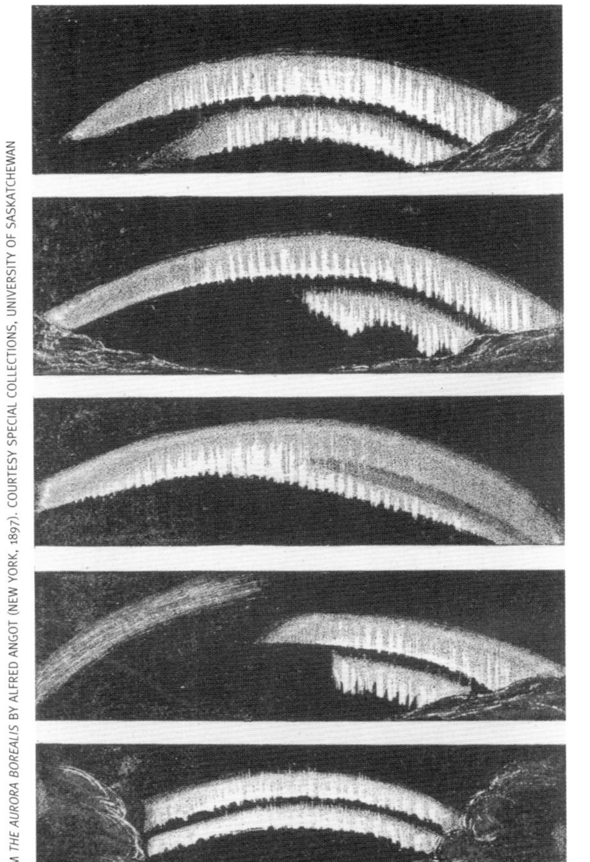

FROM *THE AURORA BOREALIS* BY ALFRED ANGOT (NEW YORK, 1897). COURTESY SPECIAL COLLECTIONS, UNIVERSITY OF SASKATCHEWAN

The series of sketches at left appeared in an 1897 book by Alfred Angot called *The Aurora Borealis*, which spends considerable time discussing techniques for calculating the altitude of the aurora. During the 18th and 19th centuries, the most popular method was triangulation: measuring the angle of the line of sight from an observer on the ground to the bottom of the aurora. But this was extremely unreliable, with estimates of the aurora's height ranging from a couple of thousand feet to hundreds or even thousands of miles.

When Captain Charles Francis Hall led the *Polaris* expedition to the North Pole in 1871, the ship's doctor and his assistant used a novel method to study the aurora, as described in this narrative of the trip: "Doctor Bessels stood outside the observatory sketching and taking notes of the rapid changes of the phenomenon. He held a string leading into the magnetic snow-house, where Mr. Bryan sat watching the magnetometer. The doctor pulled the string when changes occurred, and Mr. Bryan being thus warned noted the time and read the magnetometer. In this way, they were able to trace the effect of various combinations and movements."

The sketch below, showing an Inuit hunter making a less scientific observation, is also from the expedition's narrative, published in 1876.

COURTESY SPECIAL COLLECTIONS, UNIVERSITY OF SASKATCHEWAN

Facing the Sun

A NEW CENTURY OF AURORA SCIENCE

In 1859, the year of the great solar storm, *Scientific American* noted that "a connection between the northern lights and forces of electricity and magnetism is now fully established." Yet few scientists could bring themselves to believe that the Sun played a role in the process.

In 1892, the most influential scientist of the era, Lord Kelvin, pronounced that the "supposed connection between magnetic storms and sunspots is unreal." Just as medieval philosophers had been held back by their loyalty to Aristotle, Victorian scientists bowed before Lord Kelvin, and few had the temerity to challenge his ideas.

But the dawn of the new century was a turning point in our understanding of the aurora. New discoveries in the fields of electricity, magnetism, spectroscopy and photography all helped answer long-standing questions about the polar lights. Only when the space age arrived were scientists able to get a new perspective by viewing the aurora from above. Finally, their goal was within reach.

An unusual pink aurora appears to erupt from a hillside in a coastal village in British Columbia, near the Alaska Panhandle.

In the 1890s, physicists identified electrons for the first time and demonstrated that these negatively charged particles could be deflected by a magnetic field. This discovery inspired Norwegian scientist Kristian Birkeland to pose the following question: Is it possible that the Sun is emitting streams of electrons in the Earth's direction and that these charged particles are being steered toward the poles by the Earth's magnetic field?

Building on several earlier ideas about the aurora, Birkeland came up with an elegant model. To test the idea in the laboratory, he designed a "terrella" (Latin for "little earth"), a magnetized sphere placed inside a partial vacuum to simulate Earth (see photo on page 78). When he bombarded the terrella with electrons, rings of light appeared around the poles. Although not all Birkeland's ideas turned out to be accurate, his pioneering work on the aurora paved the way for modern physicists to unlock the secrets of the northern lights.

Ahead of His Time

KRISTIAN BIRKELAND AND THE BIRTH OF A NEW IDEA

Norwegian physicist Kristian Birkeland traveled to the remote northern reaches of his homeland to test his controversial ideas about the aurora. With most of the country situated above 60°N latitude, Norway remains one of the most reliable places on Earth to view the aurora. Here, Birkeland at work with his terrella machine in the laboratory.

Many scientists are deeply devoted to their work, but few have suffered the physical hardships that Kristian Birkeland endured to study the aurora. At the end of the 19th century, the Oslo-born scientist had come up with a revolutionary idea: that the magnificent polar lights were caused by electrons from the Sun being drawn toward the Earth's magnetic poles. But Birkeland needed hard evidence to support his theory.

To gather data, Birkeland made three expeditions to Scandinavia's Far North, beginning in 1896. He braved months of bone-chilling temperatures and gloomy darkness. On one occasion, he was pinned down by a blizzard for three weeks. Two of his colleagues died in a snowslide, and a third lost his fingers to frostbite. Birkeland did, however, collect compelling evidence that auroras are related to electrical currents following the Earth's magnetic field lines.

Birkeland performed many other lab experiments that supported his theory about a solar connection to the aurora, but he was unable to piece together a complete model. For example, he could not explain how a stream of electrons — which are negatively charged and therefore should repel one another — could hold together as it raced across millions of miles from the Sun to Earth. Birkeland's hypothesis failed to gain traction when he couldn't answer his critics. (It was later discovered that the solar wind includes both electrons and positively charged protons.)

The physicist eventually moved on to other pursuits, including developing an electromagnetic gun and a process for fixing nitrogen from the air to make fertilizer. He enjoyed considerable scientific success and was nominated for the Nobel Prize seven times.

But Birkeland had a difficult personal life, struggling with insomnia, paranoia and substance abuse. In 1914, he traveled to Egypt, and when the Great War broke out later that year, he was unable to return to Norway. After three years in Egypt, he was invited to join some scientific colleagues in Tokyo, where, at age 49, he died alone from a drug overdose.

COURTESY NORWEGIAN MUSEUM OF SCIENCE AND TECHNOLOGY

FACING PAGE: © ANTONY SPENCER/VETTA COLLECTION/ISTOCKPHOTO

In the 1920s, Carl Störmer, a colleague of Birkeland, devised a way to reliably measure the height of the aurora using photography. He took simultaneous photos from two different positions, ensuring that the images contained at least two stars whose exact positions were known. Then he was able to calculate the altitude of the aurora by measuring its relative distance from the stars in the two images. Using this technique, Störmer ascertained that the bottom edge of the aurora is typically 60 to 65 miles (100-105 km) above the ground.

Then two more pieces of the puzzle fell into place in 1925. First, scientists determined the height of the ionosphere — an electrically active region of the upper atmosphere — and their findings meshed well with Störmer's measurements. That same year, two Canadian physicists discovered that excited oxygen atoms emit greenish light in a partial vacuum and red light in an ultrahigh vacuum, explaining why the aurora's colors vary with altitude.

The most prominent auroral scientist of the 1930s was Sydney Chapman. The opinionated British mathematician and physicist was dismissive of Kristian Birkeland's work, stating that the Norwegian's "direct observational contributions to auroral knowledge were slight." Chapman's bone of contention was that it was impossible for a stream of negatively charged electrons to travel from the Sun to Earth without dissipating.

Chapman wondered whether the stream of particles from the Sun was not all electrons. Indeed, a new positively charged particle — the proton — had recently been discovered. Perhaps, Chapman reasoned, the Sun emitted both electrons and protons and that, as a whole, this matter was electrically neutral. Scientists later gave a name to this thin gas of charged particles: plasma.

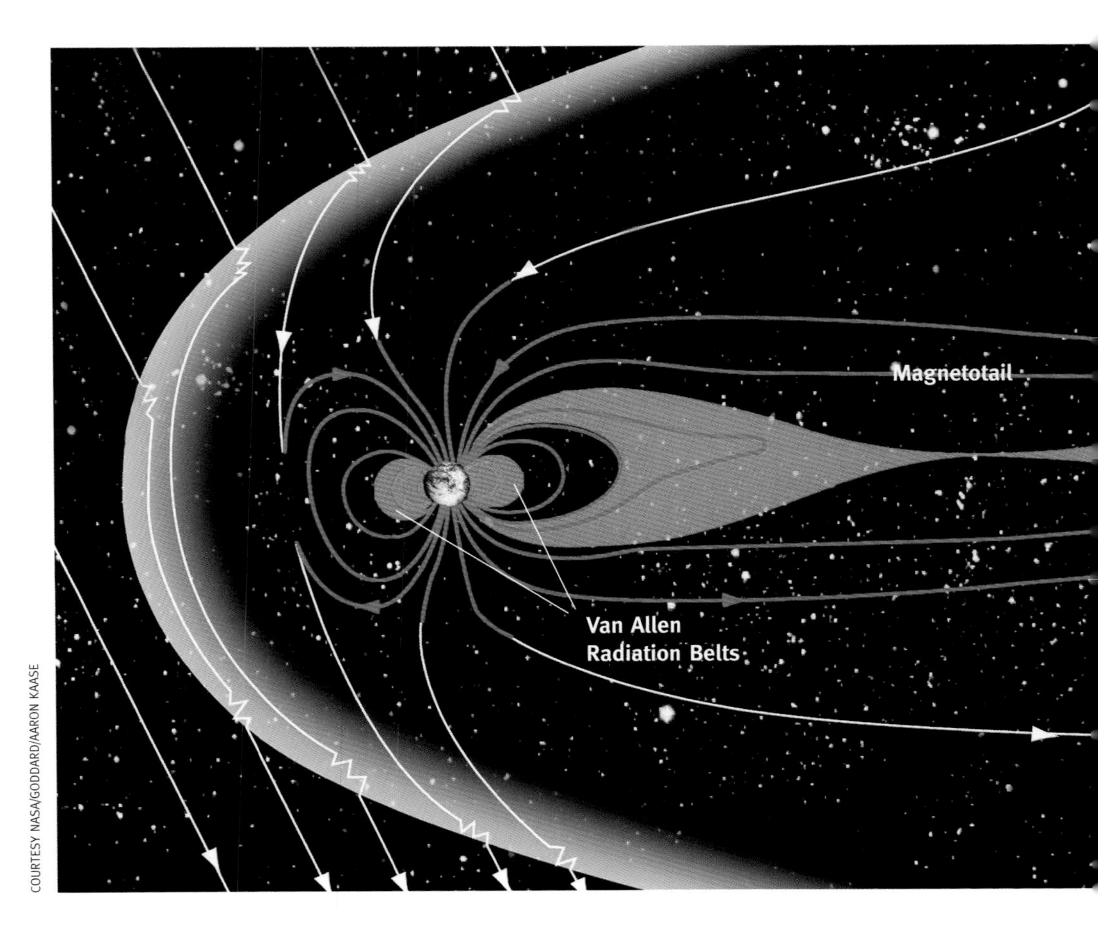

COURTESY NASA/GODDARD/AARON KAASE

As auroral science evolved in the 1940s and 1950s, more controversy arose about the physics going on in the upper atmosphere. Sydney Chapman was the reigning authority on auroras, but he was challenged by a young Swede named Hannes Alfvén. Like Kristian Birkeland before him, Alfvén was often a step ahead of his contemporaries. It didn't help that Alfvén's background was engineering, and some theoretical scientists had little respect for his ideas.

Alfvén argued that if a stream of plasma from the Sun approached Earth, it would grab the planet's magnetic field lines and drag them out into a long "magnetotail." Then, just as a current of water forms an eddy as it flows around a rock, some of the electrical current would flow back from the magnetotail toward the planet's poles. This, Alfvén claimed, was the generator that powered the aurora. There was one problem, however: The idea could not be tested with the technology of the period.

The Second World War helped speed the development of a number of new technologies, including rocketry and computers. By the early 1950s, the world's scientists had access to tools that could help them understand phenomena such as the aurora, and they decided to pool their efforts. Led by Sydney Chapman, among others, they organized the International Geophysical Year (IGY), which ran from July 1957 to December 1958 and involved some 67 countries.

During the IGY, scientists in more than a hundred locations used dome-shaped "all-sky cameras," which could photograph the sky from horizon to horizon. They coordinated their efforts so that the cameras took images at one-minute intervals, then sent their film to a central location to be analyzed. These images helped prove that the aurora does not appear uniformly all over the polar regions. Rather, the arc visible from any area is actually part of an enormous oval centered on the magnetic pole.

The scientists who organized the International Geophysical Year did not choose the dates at random. They knew that following the 11-year sunspot cycle, 1957 and 1958 would be a period of maximum activity on the Sun, and that meant a greater likelihood of brilliant auroras. They were not disappointed.

In September 1957, observers in Europe and the lower 48 states were treated to rare displays in the midlatitudes. Then, in February 1958, skies around the world erupted in a series of extraordinary crimson auroras — among the most dramatic of the century. Between these two dates, the Soviet Union launched Sputnik 1, the world's first artificial satellite; three months later, the United States followed with Explorer 1. The space age had arrived, and it ushered in a new chapter in aurora science.

The two images below, from the Solar and Heliospheric Observatory (SOHO) spacecraft, were created nine years apart and show the dramatic difference in sunspot activity around the solar minimum and maximum.

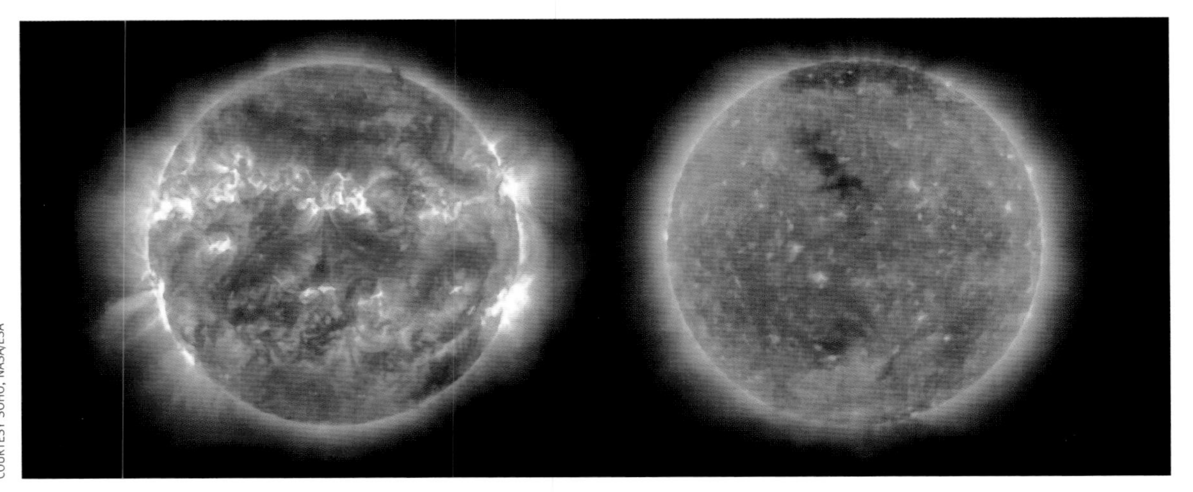

COURTESY SOHO, NASA/ESA

In 1958, Explorer 1 detected two doughnut-shaped radiation belts (see illustration on page 85) encircling Earth. These belts, named the Van Allen belts after the chief scientist of the mission, contain electrons and protons trapped by the planet's magnetic field. At the time, scientists believed that charged particles escaping from the Van Allen belts caused the aurora, but later research showed this was not the case.

During a mission to Venus in 1962, Mariner 2 confirmed that the Sun does, indeed, send streams of plasma toward Earth — the first detection of what is now called the solar wind. A dozen years later, spacecraft observed that electrical current flows from the Earth's magnetotail toward the poles, confirming Hannes Alfvén's decades-old idea, as well as vindicating a certain Norwegian aurora pioneer who had anticipated the discovery some 70 years earlier. Today, these streams of charged particles in the Earth's magnetosphere are known as Birkeland currents.

English Bay in Vancouver, British Columbia, is a popular spot to watch the sunset, and during a period of intense solar activity in 2004, the aurora also made a rare visit to this West Coast beach.

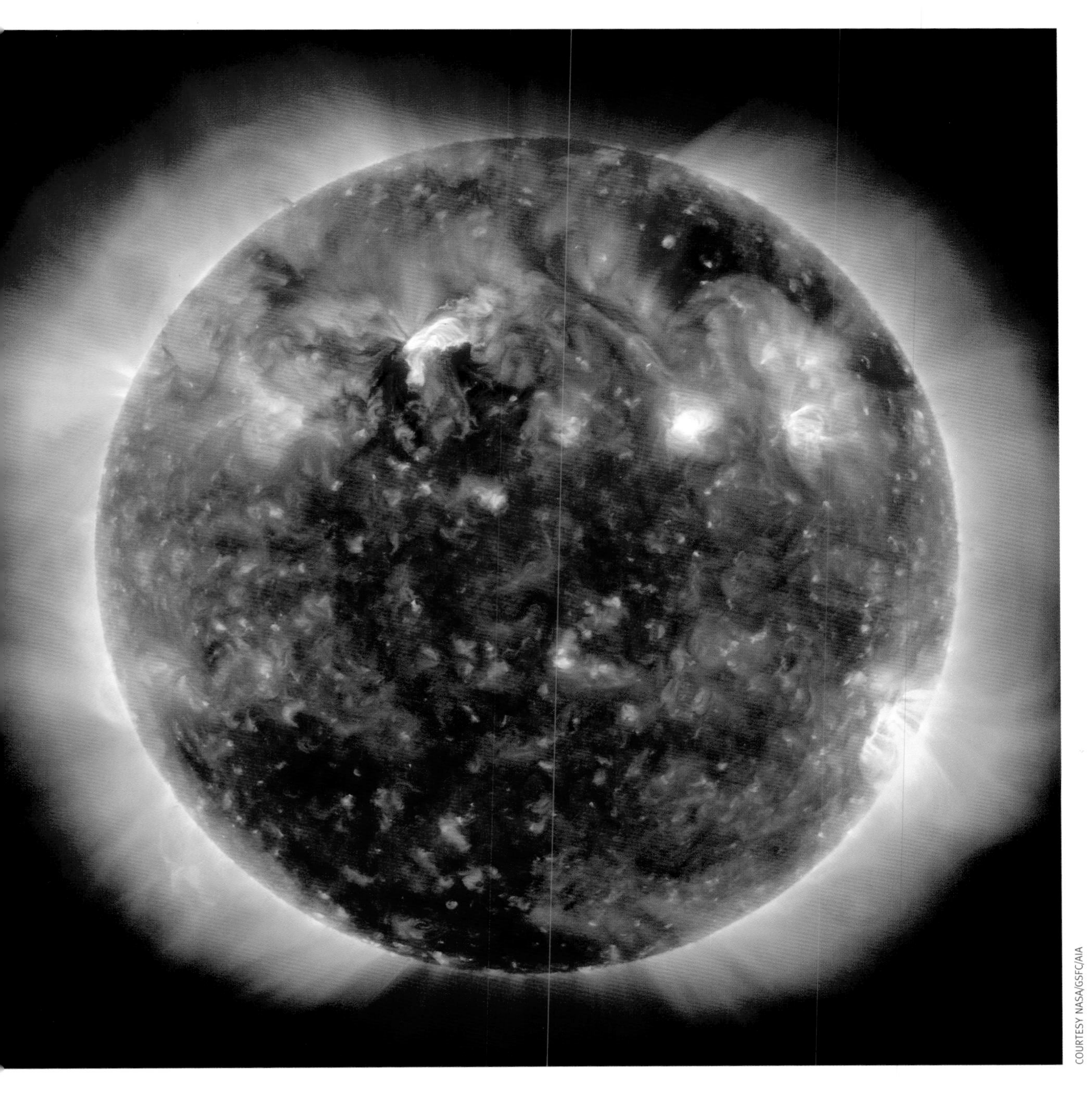

COURTESY NASA/GSFC/AIA

The number of sunspots ebbs and flows in an 11-year cycle. During periods of intense magnetic activity, large eruptions can occur around sunspot groups. These explosions cause a sudden brightening on the surface of the Sun that we call a solar flare. As English astronomer Richard Carrington first observed in 1859 — although he denied any connection — solar flares often signal an impending vivid aurora.

Sometimes, an eruption on the Sun's surface causes part of the corona — the solar atmosphere — to be blown off. Called a coronal mass ejection, this phenomenon is one of the most violent events in our solar system. It can blast billions of tons of plasma into space and send it hurtling toward Earth at 500 miles (800 km) per second. The prominence at the top right of this image is over 420,000 miles (675,000 km) across — more than 50 times the Earth's diameter.

The most intense geomagnetic storms, and the most brilliant auroras, are directly related to these solar gales.

COURTESY SOHO, NASA/ESA

A twilight aurora dances in front of a backdrop of star trails in this time exposure. Satellites have contributed enormously to our understanding of the aurora, and occasionally, they step forward to take a bow. The three bright flares at the lower right are satellites whose reflective panels have caught the glint of the Sun.

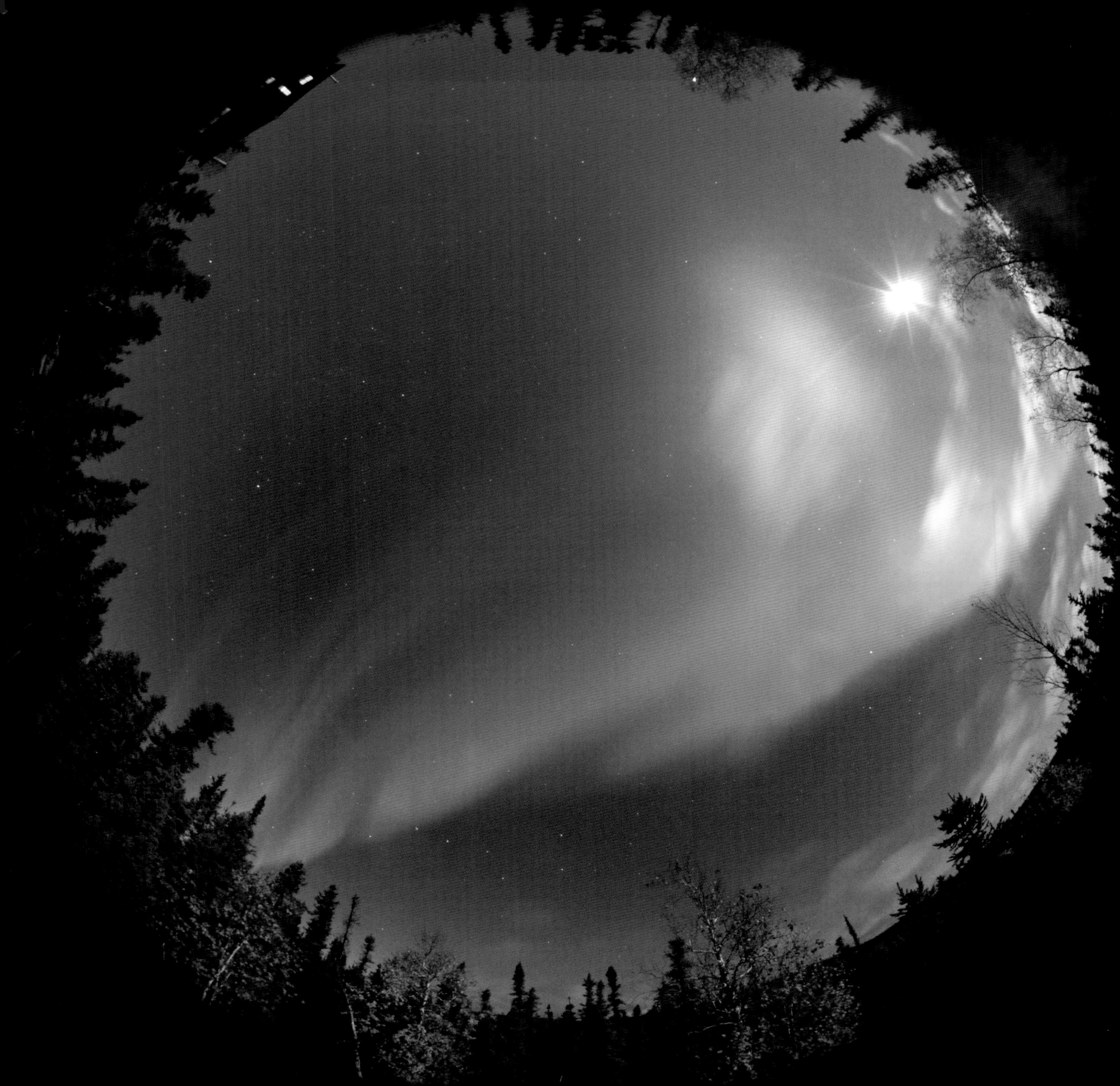

COURTESY NASA

Captain James Cook viewed the aurora australis from the deck of the *Endeavour* on September 16, 1770. At the same time, on the other side of the globe, Chinese observers witnessed a rare display of the northern lights and duly noted it in their records. When historians later made the connection between the two observations, it was the first hard evidence of a phenomenon that had long been suspected: Auroras occur simultaneously over both the north and south magnetic poles.

In 1967, scientists set out to prove that mirror-image auroras appear at the top and bottom of the world. One group flew an aircraft over Alaska at the same time that a second group flew south of New Zealand, and both recorded auroral activity with all-sky cameras. As they had predicted, the auroras followed the same patterns over the same periods. In 2005, however, data gathered via spacecraft revealed that the auroras on either side of the globe are not perfectly symmetrical.

In this series of images, the diaphanous curtains of an aurora are reflected in a small pond along the Ingraham Trail, which extends east from Yellowknife, in the Northwest Territories.

Frontiers

ON THE CUTTING EDGE OF DISCOVERY

By the 1970s, scientists had pieced together a thorough model to explain the aurora. They understood that the charged particles originate at the Sun, that the plasma streaming from the Sun interacts with the Earth's magnetic field and that the colors are caused by oxygen and nitrogen gas high in the atmosphere. But unanswered questions still remained — and many linger today.

More than a century ago, Kristian Birkeland trekked on foot into the hostile Arctic to study the aurora. Today, from unmanned probes to the International Space Station, humans carry their scientific instruments into space, where they make observations of the aurora about which Birkeland could only have dreamed.

With better tools and more resources than ever, scientists continue to improve their understanding of nature's light show. Does the aurora make a sound, as witnesses have long claimed? What causes the structures that are visible from space but not Earth? Are the auroras on other planets created by the same forces as the polar lights on Earth? For every mystery solved, another emerges.

An auroral arc ripples across a January sky as the Moon rises over the rugged Dettah Ice Road near Yellowknife, Northwest Territories.

Inuit have long claimed to hear whistling, crackling or buzzing sounds that correspond with the movement of the auroral lights. This claim doesn't seem to square with science, however. Since the bottom edge of the aurora is at least 60 miles (100 km) above the ground and sound travels about one mile (1.6 km) every five seconds, any noise would take at least five minutes to reach the listener's ear.

But not all modern scientists are ready to dismiss the idea as superstition. Sophisticated instruments have shown some relationship between geomagnetic storms and sound energy, leading one scholar at the Helsinki University of Technology to write: "The observational material supports the reality of auroral sounds."

From the perspective of an observer on the ground, an auroral band can appear to touch the horizon, forming a bridge to the skies. In the 1980s, the Dynamics Explorer 1 satellite collected data suggesting that the entire auroral oval may occasionally be spanned by a "bridge" connecting the night side of the aurora to the day side. This phenomenon is called a "theta aurora," for the Greek letter θ. Its cause is still unknown.

Until recently, the so-called pulsating aurora was one of the more baffling mysteries of the polar lights. Unlike an active display that appears to swish and sweep across the sky, a pulsating aurora blinks on and off at intervals of about 5 to 40 seconds.

In 2010, NASA scientists discovered a link between the frequency of the pulses and the activity of chorus waves — electromagnetic waves that occur thousands of miles above Earth.

A pulsating aurora is often difficult to see with the naked eye, as it is relatively dim, but a long exposure like the one at right, shot in the Nisga'a village of Gingolx, British Columbia, reveals a rainbow of colors.

An active aurora takes on the familiar spiral shape of a hurricane, but the resemblance is merely a coincidence. "Space weather" is an increasingly important field, however, and researchers are developing new techniques for forecasting magnetic events in the atmosphere and beyond. By predicting the arrival of an intense solar wind, scientists may be able to protect Earth-orbiting satellites and spacecraft from damage.

NASA's Solar Terrestrial Relations Observatory consists of a pair of probes that trace the same orbital path about the Sun as the Earth's — one is positioned in front of the planet, the other trails behind. The two probes are designed to analyze coronal mass ejections, the massive gusts of solar wind that can cause intense auroras.

At left, the characteristic whorls of an auroral breakup over Yellowknife. When an aurora is directly over the observer, the vertical rays of light appear to converge far overhead.

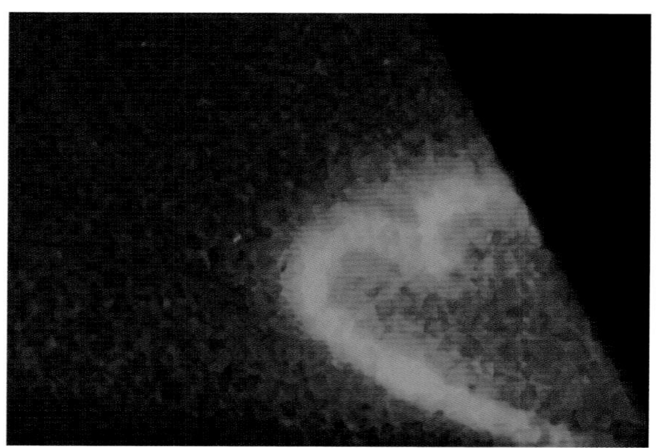

COURTESY NASA/POLAR

Launched in 1996, NASA's Polar spacecraft studied solar wind activity for more than a dozen years. The probe orbited perpendicular to the Earth's equator, passing over the north and south poles about every 18 hours. The photo above, aptly nicknamed "Broken Heart," is the last image the Polar spacecraft sent back before it was switched off in 2008.

Orbiting about 200 miles (320 km) above Earth, the International Space Station (ISS) sometimes passes over or even directly through the aurora. An astronaut on the ISS captured this photo, top right, of the aurora borealis over northern Canada. The large circle in the upper portion of the image is the 60-mile-wide (100 km) Manicouagan Reservoir, created more than 200 million years ago when an asteroid slammed into Earth. Bottom right: On the other side of the globe, an ISS astronaut snapped this image of the aurora australis in 2010.

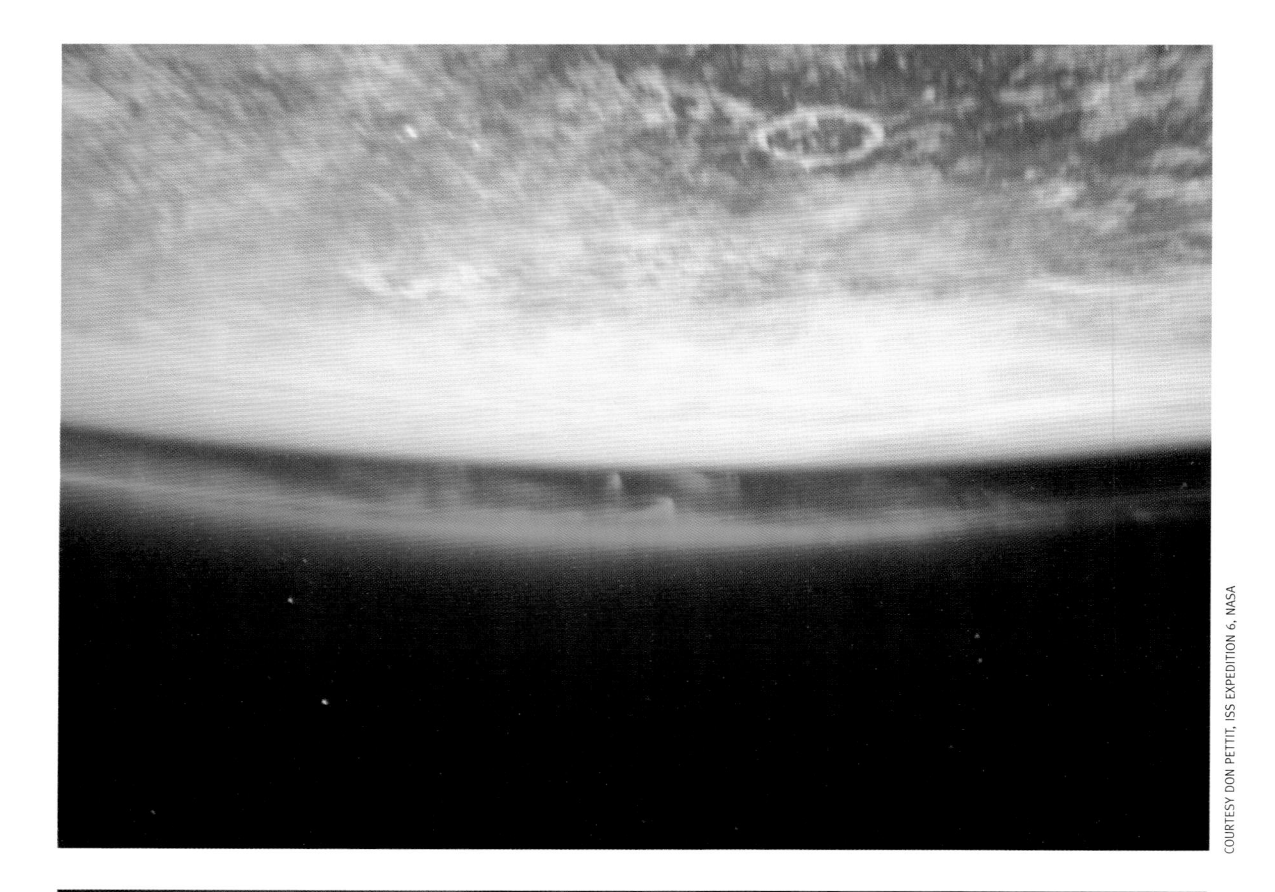

COURTESY DON PETTIT, ISS EXPEDITION 6, NASA

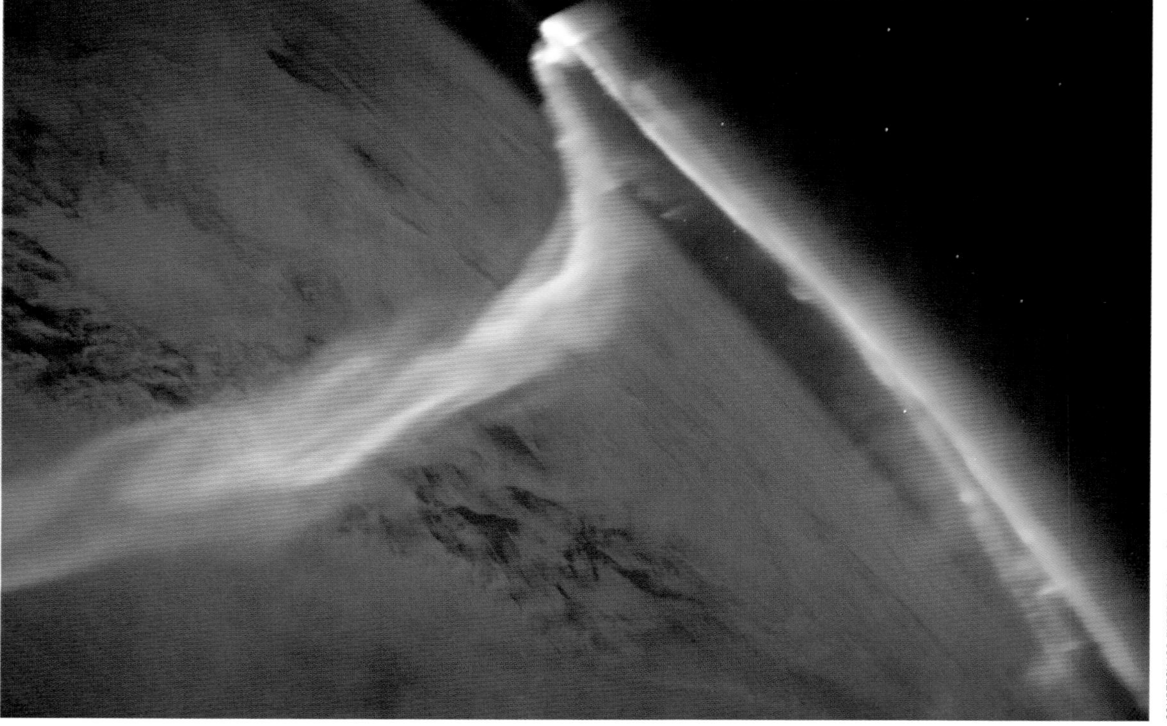

COURTESY ISS EXPEDITION 23 CREW, NASA

The landscape near Yellowknife is bathed in a wash of rainbow colors in both the aurora and its celestial backdrop. The bright pink disk at the far right in this long-exposure image is the planet Mars, rising in the east. The V-shaped group of bright stars at left forms the face of the bull in the constellation Taurus, with the bluish Pleiades, or Seven Sisters, above it and to the right.

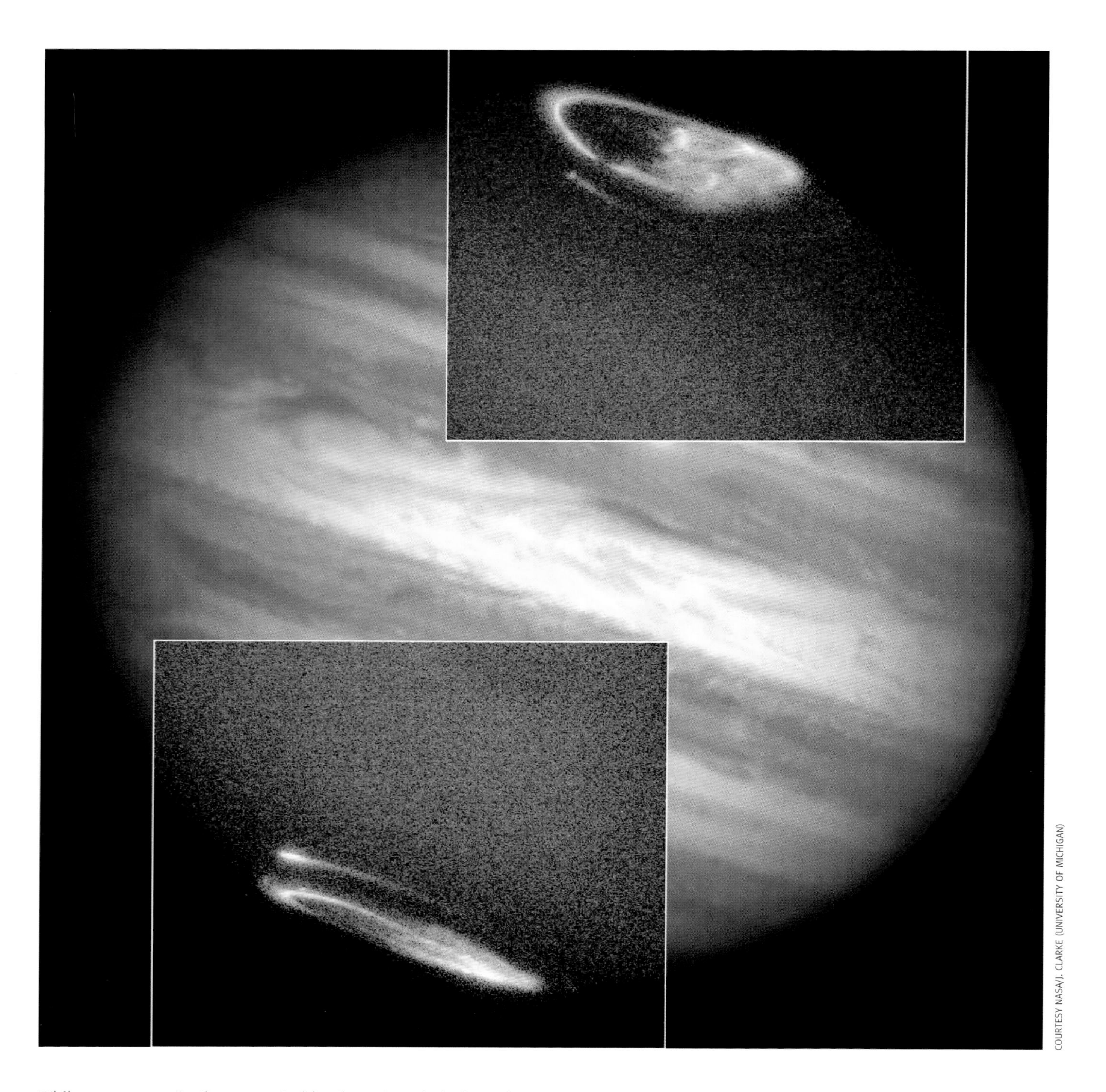

COURTESY NASA/J. CLARKE (UNIVERSITY OF MICHIGAN)

While auroras on Earth are created by the solar wind, the polar lights on Jupiter are generated by a different source. Most of the charged particles that ignite Jupiter's extremely bright auroras are emitted by the moons of the giant planet, especially the volcanic Io.

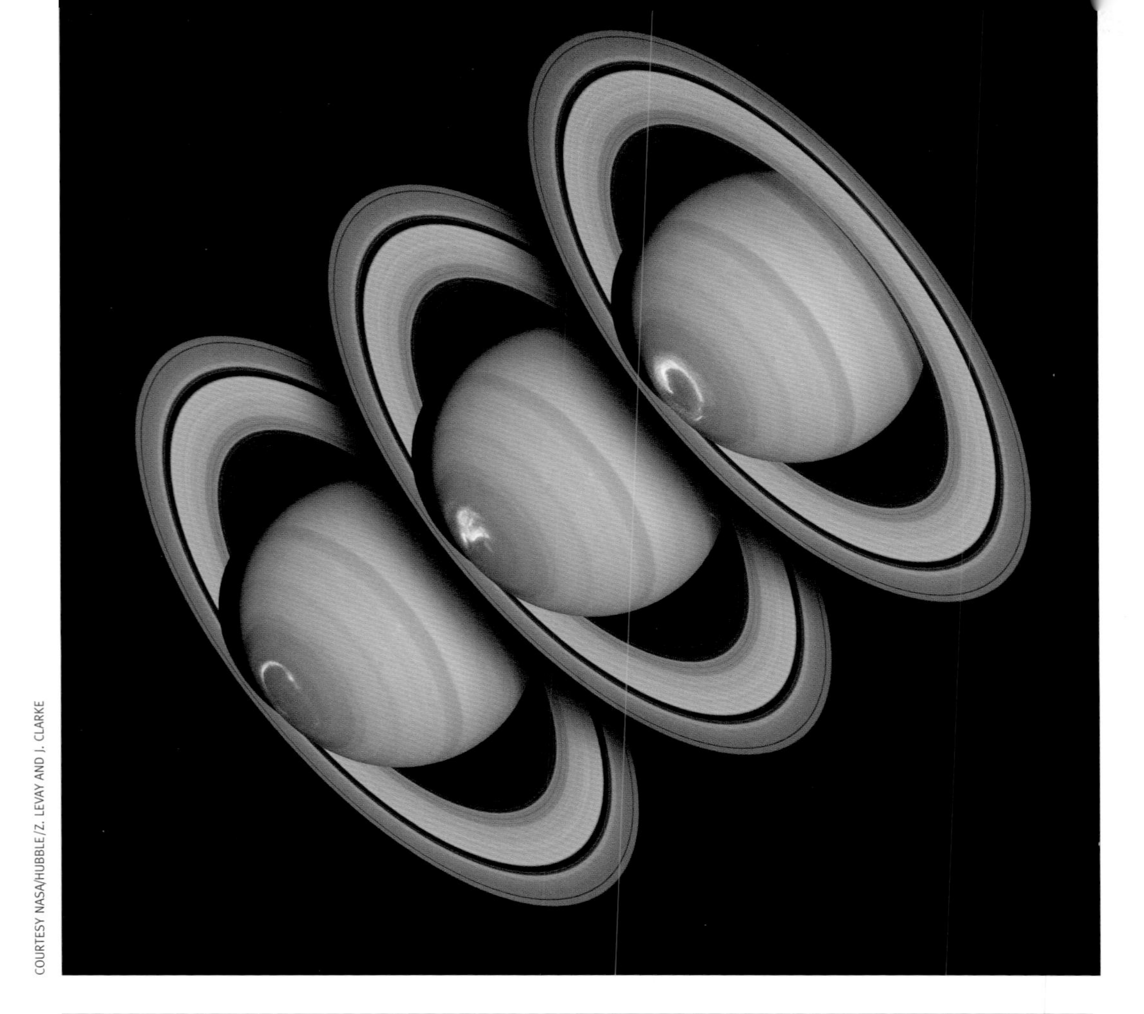

COURTESY NASA/HUBBLE/Z. LEVAY AND J. CLARKE

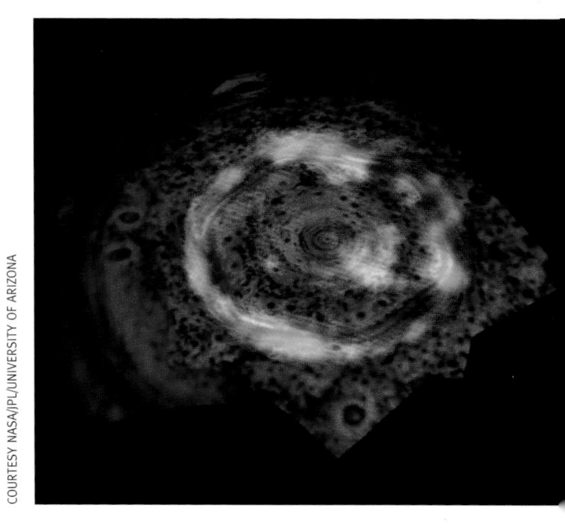

COURTESY NASA/JPL/UNIVERSITY OF ARIZONA

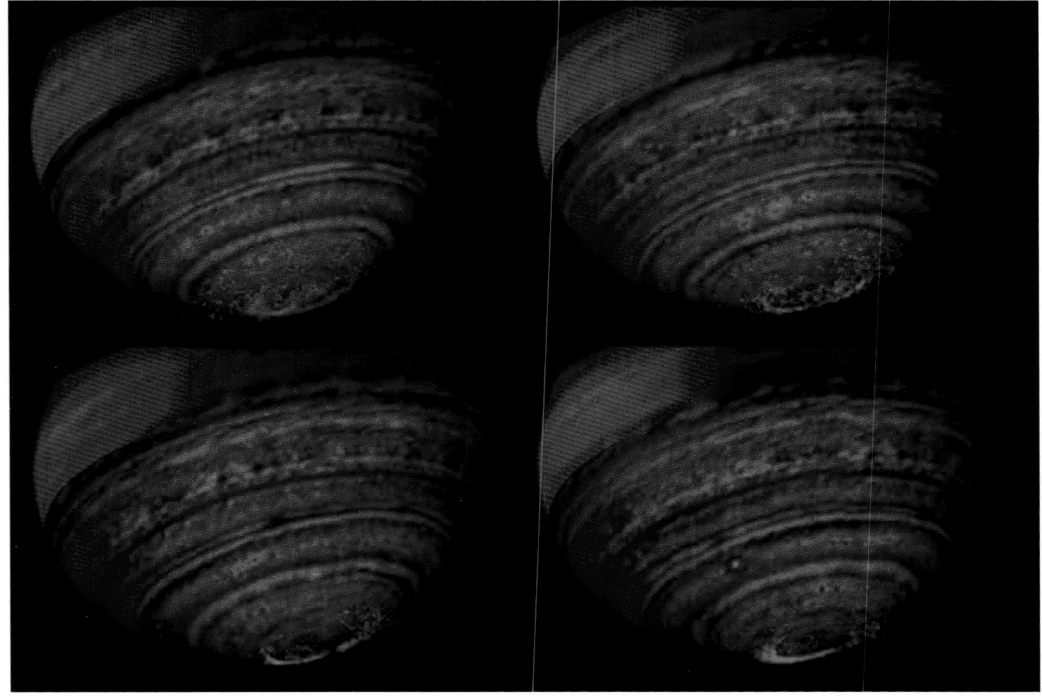

COURTESY NASA/JPL/UNIVERSITY OF ARIZONA/UNIVERSITY OF LEICESTER

Unlike those on Jupiter, the primary auroras on Saturn are fueled by the solar wind, as they are on Earth. However, when the Cassini spacecraft took an infrared image of Saturn's north magnetic pole in 2008, above, scientists were surprised by what they saw. The blue region represents a previously unknown secondary aurora blanketing a large area centered on the pole. So far, it remains unexplained.

While auroras on Earth typically last no more than a few hours, they can persist for days on Saturn, as shown in this composite of images, top left, from the Hubble Space Telescope and the Cassini spacecraft.

The changing nature of the aurora around Saturn's south magnetic pole is revealed in the four composite images (shown in false color) at bottom left. The dark spots and bands are the silhouettes of clouds and small storms illuminated from beneath. The images were created using data collected by NASA's Cassini spacecraft, which has been orbiting the ringed planet since 2004.

Keats is said to have accused Sir Isaac Newton of “unweaving the rainbow” — once the spectrum of colors was explained by science, its beauty seemed somehow diminished. Does the aurora become less wondrous when you understand that it is caused by magnetism and not magic, by gases and not goddesses? As American physicist Richard Feynman once wrote: “I don’t see how studying a flower ever detracts from its beauty. It only adds.”

Glow Getters

EXPERIENCING THE AURORA YOURSELF

The aurora has captivated Arctic travelers since the first explorers arrived in the 16th century. Today, people from all over the world come to northern communities and brave the frigid weather as they try to catch a glimpse of the famous lights. Like many of nature's most beautiful spectacles, the aurora can be fickle, and some tourists leave disappointed. But if you understand when and where an aurora is most likely to appear, you can improve your odds of being in the right place at the right time.

When you are fortunate enough to be treated to a colorful auroral display, you may want to capture it with a camera. It helps to have top-of-the-line equipment, but even amateur photographers can get memorable shots after learning a few basic techniques. While the images of professionals such as Yuichi Takasaka can reveal the sinuous structure and subtle colors of an aurora, there's no substitute for seeing the light show firsthand. "You really have to be there," says Takasaka. "It's unbelievable."

A twilight aurora appears to brush the treetops in the coastal community of New Aiyansh, British Columbia.

Although auroral displays are often associated with winter, auroras are no more frequent during this season than any other. At high latitudes, however, summer nights are extremely short, rendering the aurora almost invisible. During the long nights of the polar winter, auroras are much more likely to be seen — if you can brave the cold. Many aurora enthusiasts travel to Yellowknife, Northwest Territories, for the best views, as the auroral oval passes right over the city, located at 62°N latitude.

While the aurora is often visible within the city limits, several tour operators have set up shelters on the outskirts, away from the lights and smog. Above, an auroral band stretches over the Bush Pilots Monument in Yellowknife.

The flat land and dry air around Yellowknife make it possible to appreciate the many wonders of the sky — sometimes all at once. During this night on the shores of Pontoon Lake, just east of the city, an aurora fills the scene as a bright Moon is setting in the southwest. The glowing red beacon in the center of the image is Mars. When this photo was taken in September 2003, the red planet was at its maximum brightness, as it was making its closest approach of the century.

The solar bursts that ignite auroras can take two to three days to reach Earth, so aurora watchers often have advance warning of an impending display. One good night of viewing is often followed by another, since intense magnetic activity in the Earth's atmosphere can last for several days.

The 11-year cycle of solar activity (from one solar maximum to the next) has a dramatic impact on the intensity and frequency of auroras. While displays can be seen at any time, they are most common for a couple of years before and after solar maximum. In addition, the Sun rotates on its axis about every 27 days. Because auroras are related to active sunspots that face Earth, particularly good displays may be seen approximately 27 days apart.

Keeping Vigil

TIPS FROM A VETERAN AURORA WATCHER

Aurora photographers set up their tripods and cameras on the shores of Great Slave Lake, in Canada's Northwest Territories, to capture a mid-September light show. The path of moonlight reflected on the water is an added bonus.

Yuichi Takasaka saw his first aurora in Jasper, Alberta, in 1990. "I was really disappointed," recalls the professional photographer, who moved to Canada from his native Japan as a teenager. "It was just like a little green cloud. I forgot all about it." When Takasaka relocated to Yellowknife a few years later, however, he was captivated by the displays of aurora borealis that lit the night skies over the Northwest Territories. Since then, he has regularly led groups of tourists on excursions to observe the aurora from this ideal location.

If you're thinking about experiencing and photographing the aurora yourself, Takasaka has some suggestions. While there is no seasonal peak in activity, he prefers to go at the end of summer.

"My favorite time to observe auroras, at least in Yellowknife, is the first three weeks of September," he says. "By then, the mosquitoes are gone, but it's still not too cold — the temperature hovers around freezing. The lakes are not frozen yet, so you get the reflection of the aurora on the water, which is very nice. That is what a photographer wants to see."

In Yellowknife, the aurora can be visible 200 nights a year or more. Although it may appear whenever it's dark, Takasaka explains that the best time for viewing is "magnetic midnight," when the magnetic north pole is directly between you and the Sun. "Usually, I recommend observing right around actual midnight — maybe about two hours before until two hours after."

While you can see an active aurora from a city, you'll get the best views if you observe from a dark site away from any light pollution. Takasaka uses a cabin about a half-hour outside of Yellowknife. When driving to a remote location in the cold and dark, put safety first. Takasaka has taken photos in temperatures approaching 40 below, so he has special boots and clothing to protect him from extreme cold. "I'm not so concerned about myself — it's more about the car," he jokes. He brings an extra battery and starts the engine every two hours to warm up the oil.

When a particularly intense magnetic storm erupts, the auroral band expands to the south. "If you are in Yellowknife, the belt goes right past you," says Takasaka, "so there is nothing to see." In July 2004, he anticipated a powerful aurora and hopped a plane to Vancouver, where he was able to photograph it downtown. Then he flew back to Yellowknife for the encore. "The energy always has to return. It expands to the south and then comes back, so you can see it twice."

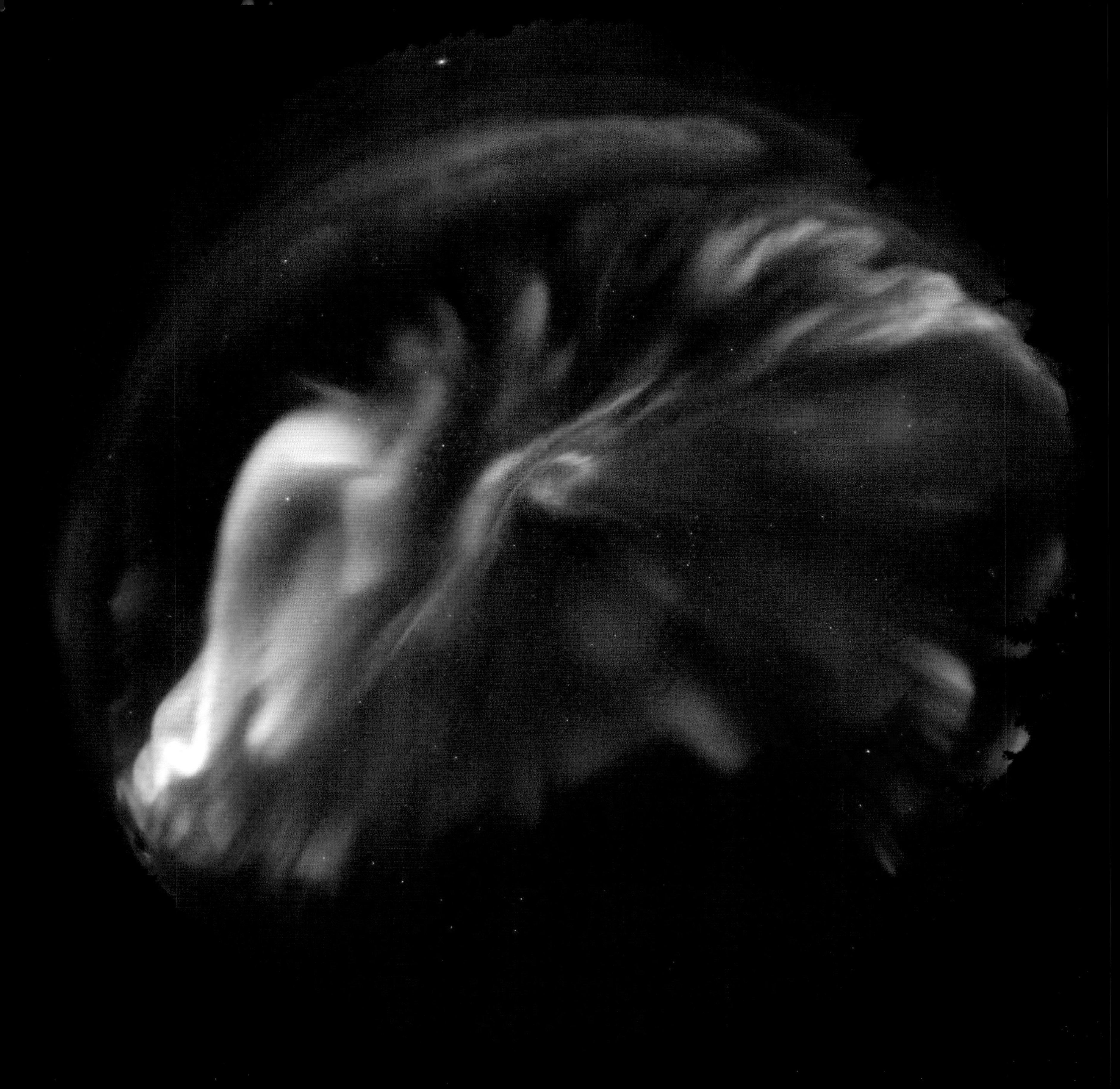

The luckiest aurora viewers get to experience what's called an auroral breakup, seen above and on the facing page. "I guide a lot of TV and magazine crews from all over the world," says Takasaka, "and when they see an aurora breakup, they just go nuts. Sometimes, you get really nice colors — the pink and purple are just great."

The breakup occurs when the aurora moves directly over the observer and the vertical rays converge far overhead. "If you're downtown in a big city, for example, and you look up, all the buildings seem to be tipped toward the center," explains Takasaka. "It's the same thing with an aurora breakup. It looks as if you are totally surrounded by lights that are hundreds of miles tall." Along the hem of the auroral curtain, nitrogen glows pink and purple. "The bands are always moving, sometimes very quickly. You can see them swirling around, as if a bomb just went off."

Cameras are more sensitive than our eyes in detecting light at the red end of the spectrum. As a result, photographs, especially long exposures, show reddish hues that are not visible to the naked eye.

Streaks of red may appear during a typical auroral display, especially at high altitude, but during a powerful magnetic storm, the entire sky turns crimson. All-red auroras frightened medieval Europeans, and even today, some urban dwellers have been known to call the police or the fire department when they witness one. But for Takasaka, they're a reward for spending hundreds of hopeful nights in the cold. "Every photographer wants to take an image of a red aurora. The first time I saw one, I was jumping up and down with excitement."

"I like to photograph auroras with something else in the frame," says Yuichi Takasaka, who took these two images at Lady Evelyn Falls on the Kakisa River, in the Northwest Territories. To shoot a still image, Takasaka suggests starting with an exposure of about 20 seconds, then checking the digital image and adjusting the time up or down as necessary. To create an image with star trails, as above, Takasaka mounts his digital camera on a sturdy tripod and programs it to take a series of short exposures. Then he "stacks," or combines, the images using special computer software.

The auroral band can snake across the sky from horizon to horizon. The extent of its reach can be seen through an 8mm 360-degree fish-eye lens, which captures an image of the entire visible sky, above. Right: By digitally stacking hundreds of images taken at timed intervals, Takasaka creates long, concentric star trails as Earth rotates. The stationary star at the center is Polaris.

Some of Takasaka's favorite aurora images are those that capture a meteor blazing across the field of view. By the time it begins to glow, the meteor is often at an altitude of less than 60 miles (100 km), just below the bottom edge of an auroral curtain.

An aurora that occurs near dusk or dawn can take on a bluish green color. Takasaka took this image, facing page, at twilight at Prelude Lake, near Yellowknife.

Sun Spotting

PLANNING YOUR AURORA WATCHING

If you live in a northern community or plan to visit one, a little advance preparation can help you get the most out of your aurora watching. Yuichi Takasaka uses astronomy software that can generate an image of the sky for any date, time and location. The program shows when the Sun, Moon and planets will be above the horizon, which helps him imagine backdrops. For example, Takasaka likes to include the rising or setting Moon in many of his shots.

Even though the aurora is not as predictable as the motion of the stars, it is possible to get advance warning that one may be in the offing. Activity in the Earth's magnetic field is measured on a scale called the Kp index, which ranges from 0 to 9 (6 or more indicates a geomagnetic storm that may give rise to an aurora). The index is updated every three hours on several websites, including those listed here.

In this image, taken near Whitehorse, Yukon, Jupiter is low on the horizon to the left of center and can be seen reflected in the lake. Above and to the right are the bright disk of Saturn and the V-shaped portion of the constellation Taurus.

SITES OF INTEREST

NOAA: Space Weather Prediction Center
www.swpc.noaa.gov/Aurora/
Maps and tables help you determine the possibility of seeing an aurora in your area.

Space Weather Canada
www.spaceweather.gc.ca
Forecasts of solar activity for all of Canada and projections for the likelihood of a visible aurora.

Geophysical Institute at the University of Alaska Fairbanks
www.gedds.alaska.edu/AuroraForecast/
Regularly updated forecasts for Alaska and Yukon, including an e-mail alert service.

University of Alberta AuroraWatch
corona-gw.phys.ualberta.ca/AuroraWatch
Detailed information on auroral conditions centered around Edmonton, including e-mail alerts.

SELECT BIBLIOGRAPHY

Akasofu, Syun-Ichi. *The Northern Lights: Secrets of the Aurora Borealis*, revised edition. Anchorage: Alaska Northwest Books, 2009.

Akasofu, Syun-Ichi. *Exploring the Secrets of the Aurora*, 2nd edition. New York: Springer, 2007.

Bone, Neil. *Aurora: Observing and Recording Nature's Spectacular Light Show*. London: Springer, 2007.

Eather, Robert H. *Majestic Lights: The Aurora in Science, History, and the Arts*. Washington. D.C.: American Geophysical Union, 1980.

Freeman, John W. *Storms in Space*. Cambridge: Cambridge University Press, 2001.

Hall, Calvin, and Daryl Pederson. *Northern Lights: The Science, Myth, and Wonder of Aurora Borealis*. Seattle: Sasquatch Books, 2002.

Jago, Lucy. *The Northern Lights: The True Story of the Man Who Unlocked the Secrets of the Aurora Borealis*. London: Hamish Hamilton, 2001.

Savage, Candace. *Aurora: The Mysterious Northern Lights*. Vancouver and Toronto: Douglas & McIntyre, 1994.

INDEX